ADHD AND NEUROFEEDBACK EXPLAINED

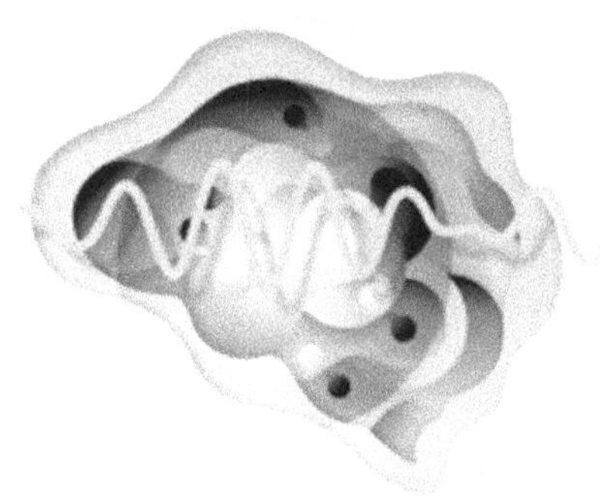

ALSO BY SCHUYLER CUNNINGHAM

Book Chapter: Treatment Planning for
Attention Deficit Hyperactivity Disorder.
Social Workers' Desk Reference. Oxford University Press.

Book Chapter: Neurofeedback:
An evidenced based non-medication intervention.
Social Workers' Desk Reference. Oxford University Press.

ALSO BY TIM NORMAN

Helping Your Husband with ADHD:
Supportive Solutions for Adult ADD/ADHD

ADHD AND NEUROFEEDBACK EXPLAINED

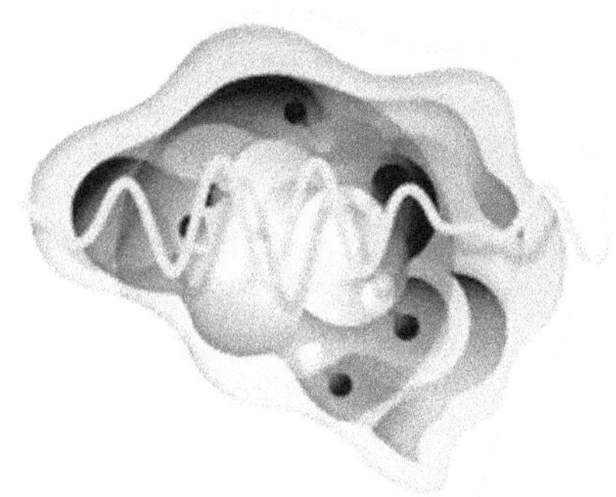

SCHUYLER C. CUNNINGHAM,
LICSW, LCSW, LCSW-C, BCD
and TIM NORMAN, LCSW, M.Ed.

Pictures of Johnny Head-in-Air were used from Heinrich Hoffmann's publication of *Struwwelpeter* (1845). For English translations of these stories, please see *The Project Gutenberg EBook of Struwwelpeter: Merry Tales and Funny Pictures* by Heinrich Hoffman.

Cover design by Ivica Jandrijevic
Interior layout and design by www.writingnights.org
Book preparation by Chad Robertson
Edited by Carmen Smith

ISBN: 9798479028014

The authors have kept the information as up-to-date as possible up until the date of publication. However, the authors recognize that current technologies, organizations, business, etc. references may have changed or no longer be in existence.

The information presented in this book is for educational purposes only. Neither the authors, nor anyone associated with this book or the information herein, claim that it is a substitute for professional medical and/or mental health care. The authors, publishers, and/or anyone associated with this book, website, or entity under the name of this book are not responsible or liable for any loss or damage allegedly arising from any information, recommendations, suggestions, or similar in this book. Our goal is simply to educate our readers about neurofeedback and ADHD and how neurofeedback is one of many treatments for ADHD. We strongly recommend that you seek professional medical and mental health care for any condition or symptoms of any condition you or those you know experience.

Printed in the United States of America.
Printed on acid-free paper.

24 23 22 21 8 7 6 5 4 3 2 1

CONTENTS

TABLE OF FIGURES

PREFACE

By Steven Warner, PhD, BCN, Q-EEGD

It is with great pleasure that I was asked to write a preface to this short treatise on neurofeedback (NFB) for ADHD. Having practiced neurofeedback for 10 years, taught nationally, mentored numerous enthused professionals in the processes of effective NFB, I can state unequivocally that this booklet fills a void that currently exists in this field. As the authors correctly state, there are numerous professional texts, research reports, and professional meetings for the practitioners of our skill, but there is a paucity of clear explanations for the lay public.

This booklet succinctly describes the history of attentional disorders and the controversies that still exist in adequately explaining and thus treating ADHD. Mr. Cunningham and Mr. Norman, both licensed mental health professionals with extensive experience in this field, cogently lay out the defining characteristics of this difficulty and the outcome of insufficient treatment to one's future success and self-esteem.

They describe the current conventional state of the art in treatment of ADHD through medication management, which carries serious risks in terms of side effects both emotionally and physiologically. Their description of the use of EEG is spot on, and then they show how we convert the raw brain signals into normative brain maps. This allows us to identify dysregulation in the brain's circuitry to understand the specifics of an individual's difficulties more accurately.

This step leads to greater specificity in neurofeedback choices for intervention and to also gauge our progress from pre-treatment to post-treatment. This objective process provides opportunity to correct less effective protocols for greater efficiency and response.

I am pleased with the detail and amount of relevant research the authors provide in the booklet. Questions about relevant research are commonly asked by professionals, parents and adults seeking help. This booklet shows how neurofeedback is definitively an evidence-based intervention. The section that compares prescribed medication to neurofeedback, a natural intervention, is necessary for assisting potential clients in deciding which route might be best for them.

Another significant issue frequently raised by clients and professionals is cost. Mr. Cunningham and Mr. Norman clearly delineate the myths and misconceptions about the expense of neurofeedback, showing how it is actually highly cost-effective, reasonably short-term, and long-lasting. This addresses the misunderstanding that may keep the public from seeking neurofeedback or referrals from medical and mental health professionals.

The authors also offer the alternative of training from home, what we refer to as remote training. More and more people are choosing this option for a variety of reasons and obtaining desirable outcomes under the care and supervision of their provider.

Finally, the authors briefly discuss the procedure of neurofeedback: how it works, what one does, what happens during a training session, and what to expect. Having this preparatory understanding tends to allow clients to begin their neurofeedback training with greater comprehension and engagement.

In summary, this is a booklet that I will provide to my clients and referral services and kudos to Mr. Cunningham and Mr. Norman for not only filling this void, but doing so in such a warm, caring, and honest manner.

Steven Warner, PhD, BCN, QEEG-D
Clinical Director, Stress Therapy Solutions
Private Practice specializing in Neurofeedback

WHY WE WROTE THIS BOOK

Neurofeedback training is a highly effective, medication-free alternative for ADHD treatment, but it is too often presented as dense neuroscience and bioengineering jargon, which makes the literature on neurofeedback inaccessible to the lay reader. Therefore, the primary goal of this book is to provide an accessible description of neurofeedback and the powerful improvements it can facilitate, especially for people with ADHD. Our goal as mental health professionals is to help you to feel smarter and more informed on the subject of neurofeedback.

We recognize that most neurofeedback materials tend to cater to neurofeedback practitioners, instead of those who will actually receive treatment or refer others for treatment. We hope there is something for everyone in this book—from people who know nothing about neurofeedback and the eager client seeking relief, to seasoned clinical professionals.

We have observed that every brain has a spectrum of symptoms associated with mental illness. In some cases, medications work well to inhibit symptoms, and in other cases talk therapy works well. We have found, and want to share with you, that there is a growing number of people for whom neurofeedback works better than any of the other treatments for ADHD. The medical field has yet to provide a clear explanation why one, or a combination, of talk therapy and/or medication treatments works better for some and not others. Instead of focusing on answering that question, we will focus on how and why neurofeedback works, as we come to better understand our brains, ADHD, and how to reduce/eliminate symptoms.

This is not a comprehensive explanation of ADHD, mental health, or neurofeedback. The field of neurofeedback is rapidly growing and evolving. The information in this book represents our points of view from our clinical training and experience treating ADHD using neurofeedback. We hope that you will find this book to be a useful tool, whether you are a referring physician, a parent of a child with ADHD, an individual currently receiving mental health treatment or thinking about getting help, or someone merely curious about ADHD and neurofeedback. More than anything, we want you to feel empowered with data and information so that you can make the most informed decision for yourself, your child, or your loved one dealing with ADHD.

We wish you all the best!

Schuyler Cunningham, LICSW, LCSW-C, LCSW, BCD
EMDR Consultant
Director, Center for Neurocognitive Excellence, LLC
www.adhd-center-dc.com
+ 1 202-998-ADHD (2343)

Tim Norman, LCSW, M.Ed
Founder and Clinical Director
The ATTN Center, New York, NY
www.ATTNcenter.nyc

CHAPTER 1

HOW NEUROSCIENCE CONCEPTUALIZES ADHD

ADHD stands for Attention-Deficit/Hyperactivity Disorder. The term ADHD is relatively new to the field of psychology and first appeared in the Diagnostic and Statistical Manual of Mental Disorders, fourth edition (DSM-IV, 1994). The DSM is the manual that describes and defines all known mental health conditions. ADHD is included, and updated, in the DSM–5, which is the current version used in the field. However, the symptoms we currently observe in those with ADHD were documented in medicine over 200 years ago.

Focused, scientific observation of people with symptoms we now call ADHD took place long before the DSM was published. While many of these records have been lost, some remain. In 1798, Sir Alexander Crichton, a Scottish physician, became the first known person to document traits similar to

ADHD. Crichton worked to draw the line between typical distraction and distraction that digressed into something that interfered with optimal functioning. Crichton found that those with symptoms comparative to what we now call ADHD struggled to remain focused. There were too many external and internal stimuli vying for attention—from the barking dog outside, the changes in room temperature, and the conversations in the background, to the clattering of horse hooves outside. Those with attention difficulties struggled to function because they couldn't filter out stimuli that, for many others, was second nature. Crichton (1798) described this attention challenge as "the incapacity of attending with a necessary degree of constancy to any one object" (reprint p. 203). This depiction is consistent with the American Psychiatric Association's description of ADHD established in 2000.

Another significant moment came in 1844 from Heinrich Hoffmann, a German physician. His notoriety was brought about by an unlikely source: an illustrated children's book he made as a Christmas gift for his son, which eventually enjoyed worldwide recognition. Two illustrated stories in the book are still discussed today in ADHD circles because they so aptly describe the condition, especially the most popular one called "Fidgety Phil." The story of Fidgety Phil describes the life of a young boy struggling to sit still and follow the directions of his parents.

Dr. Hoffmann zooms in on family mealtime. This YouTube video of Fidgety Phil,[1] directed and edited by Sam Lasseter of the Sonoma Academy, does a good job of showing the ongoing frustration of both Phil and his parents while maintaining the originality of Dr. Hoffmann's story.

---

[1] https://www.youtube.com/watch?v=LqmGK_NwGH4

Dr. Hoffmann's other tale, known as "Johnny Head-In-Air," is a humorous allegory of a boy named Johnny struggling to pay attention to his surroundings. This eventually leads to him falling into a river and needing to be rescued.

Figure 1 – Johnny Head-In-Air

At first glance these tales may seem superficial. But there's a lot of meaning hidden in them. Not only do they describe some of the characteristics of modern ADHD, but they also reveal some of its social implications. In both stories, the main character is a boy with attention difficulties who experiences serious consequences because of inattentiveness. For instance, when Johnny falls into the river, the story says he'll never forget the event, indicating he feels foolish. The river fish laugh at Johnny when he falls in the water, symbolizing the social embarrassment that often accompanies inattentiveness. After creating these stories, Dr. Hoffmann later went on to be a celebrated psychiatrist.

Interference into the quality of life of those with ADHD would eventually become a major motivator for change. Soon, simple observation and stories wouldn't be enough. Meaningful ADHD treatment forms would need to be discovered so that those with ADHD could improve their quality of life.

Skipping ahead to the contemporary understanding of ADHD, the DSM conceptualizes ADHD differently than how neuroscience conceptualizes ADHD, and these differences in diagnostic criteria lend valuable insight into the disorder. This difference is vital to understanding the way professionals currently think about ADHD. Diagnosing ADHD according to the DSM–5 uses clusters of behavioral symptoms that can be observed and reported. In other words, to diagnose ADHD we have to determine that the client has the requisite symptoms and number of symptoms as laid out in the DSM, that the symptoms impair social functioning, and that there is no better explanation for the symptoms, such as anxiety and/or depression.

For example, to make a diagnosis consistent with the DSM–5 a clinician is required to:

Observe six or more of the following symptoms to be clinically significant in children up to 16 years old and five or more for people 17 or older; observe that these symptoms were present for at least six months in such a manner as to disrupt the person's developmental stage. The symptoms include:

1. Often fidgets with or taps hands or feet, or squirms in seat
2. Often leaves seat in situations when remaining seated is expected
3. Often runs about or climbs in situations where it is not appropriate (adolescents or adults may be limited to feeling restless)
4. Often unable to play or take part in leisure activities quietly
5. Is often "on the go" acting as if "driven by a motor"
6. Often talks excessively
7. Often blurts out an answer before a question has been completed
8. Often has trouble waiting their turn

9. Often interrupts or intrudes on others (e.g. butts into conversations or games). (American Psychiatric Association, 2013)

As you can see, the DSM–5 does not include any criteria related to brainwave activity, called electroencephalogram (EEG), in the diagnosis of ADHD. At present, to obtain a reliable ADHD diagnosis, no EEG is necessary; instead, a robust battery of mental health questionnaires and an in-depth conversation with the client—and/or with those who know the client well—should be utilized. This is how both of us currently assess for ADHD in our practices.

Among neurofeedback practitioners there are certain quantitative electroencephalogram (qEEG) readings that are considered to reflect brain activity consistent with someone with ADHD. These are typically referred to as phenotypes. While the data is still rapidly developing for qEEG phenotypes for ADHD, Johnstone, Gunkelman, and Lunt (2005) described patterns of EEG data consistent with symptoms of ADHD. They reported that the frontal lobe was a key area of focus for people experiencing inattention. In less than 10 percent of ADHD cases, they observed a burst of beta waves over a longer time than typical qEEG readings.

Shortly after their article was published, Arns, Gunkelman, Breteler, and Spronk (2008) shed further light on these findings. Children with ADHD who participated in their study more often showed increases in power of the theta wave band in the frontal regions of the brain. They suggest that a slow alpha peak frequency might be a better way to understand the difference between people with ADHD and those without. Their research was not conclusive enough to suggest replacing qEEG findings with DSM–5 criteria, but it is important to consider when thinking about how qEEG-guided neurofeedback shapes our thinking when treating symptoms of ADHD.

Rather than competing with the DSM–5 criteria for ADHD, the developing input from qEEG data pushes the mental health field toward better techniques for identifying conditions geared toward treatment.

CHAPTER 2

WHAT IS NEUROFEEDBACK?

Simply stated, neurofeedback (NFB) training is a way to proactively and directly train brainwave activity. This training allows brainwaves to change their activity so that they are more reflective of a brain free of the unwanted patterns. These "normal" brainwave patterns are defined by large datasets of brainwave activity. These include "controls" of people without mental health diagnoses and those who have ADHD. When your brain is compared to a "normal" brain without ADHD, it becomes clearer what is unique about your brain and ADHD. NFB training is oriented to make your brain more "normal" with the goal to eliminate the symptoms of ADHD as compared to those who don't have them. While it sounds like we are setting clients up to inform them their brains "aren't normal," it is important to remember that your brain will always be normal for you and it is our job to translate the scientific literature so that it makes sense to you and for your brain.

Figure 2 – Electro-Cap

The International Society for Neurofeedback & Research (ISNR)[2] defines NFB as using "monitoring devices to provide moment-to-moment information to an individual on the state of their physiological functioning." Above you see one example of a monitoring device known as an electro-cap.

NFB is a form of biofeedback and is sometimes referred to as brain-biofeedback. Biofeedback is defined by the Biofeedback Certification International Alliance (BCIA) as "a process that enables an individual to learn how to change physiological activity for the purposes of improving health and performance."

Unlike other types of biofeedback, NFB targets your central nervous system, specifically cortical brain waves. NFB provides your brain with spontaneous feedback. This feedback targets brain waves that are too strong or too weak, which can be the source of unwanted symptoms.

---

[2] https://isnr.org/

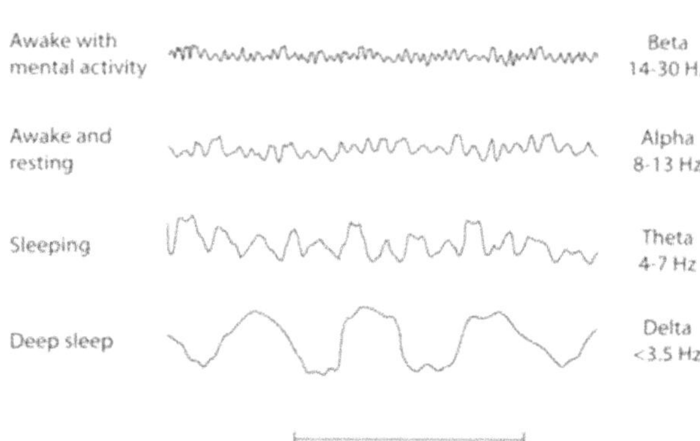

Figure 3 – Normal Adult Brain Waves

NFB may be guided by a quantitative electroencephalogram or qEEG. A qEEG is an assessment tool that allows clinicians to "see" your brainwave activity as represented by the electrical activity. The EEG brainwaves are collected by an electrode cap.

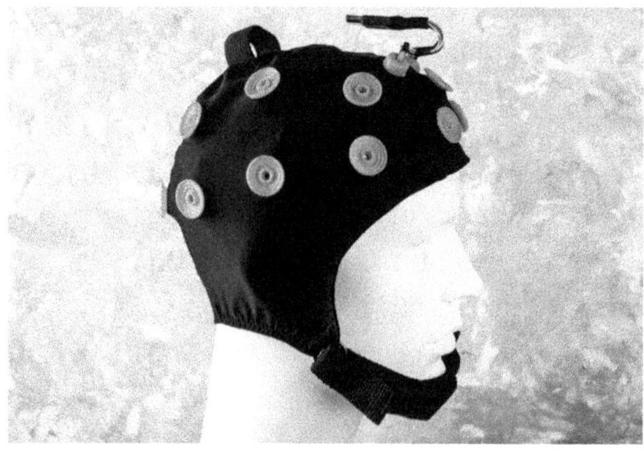

Figure 4 – Electro-Cap

Each of the electrode sensors "reads" a different location or "site." Figure 5 shows what the electrical activity ("raw waves") of a brain looks like when collected with the cap.

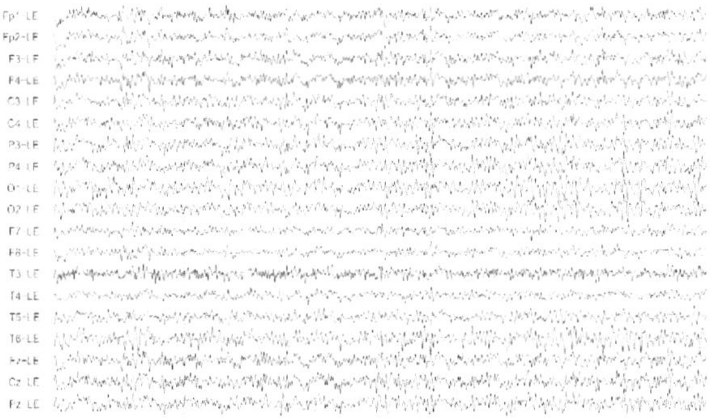

Figure 5 – Sample of raw EEG collection

Forty-two areas of the brain have been mapped and labeled for the most general description of brain regions within NFB. These areas are known as Brodmann areas. To obtain data from the Brodmann areas, we use the 10-20 International System (Figure 6), incorporated with the Electro-Cap.

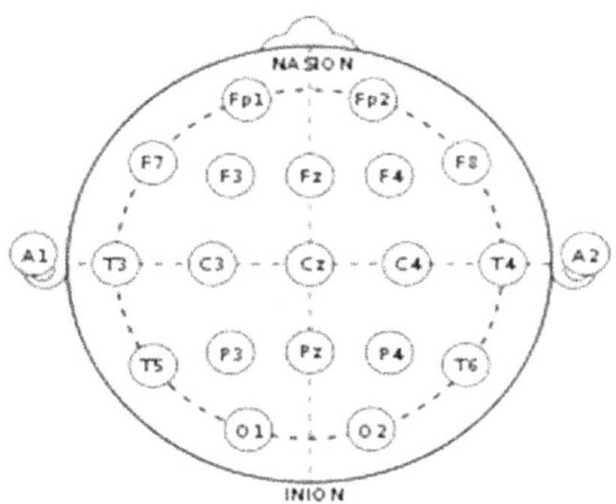

Figure 6 – 10-20 Sites EEG

Once collected, all the squiggly lines, or raw EEG as seen in Figure 5, are translated into an easier to understand brainmap (Figure 7). The brainmap is a graphical representation of brainwave activity. This brainmap is a useful way for clinicians to share your specific data with you, as well as to develop a NFB-training protocol specific to your brain.

Summary of the Z-score analyses

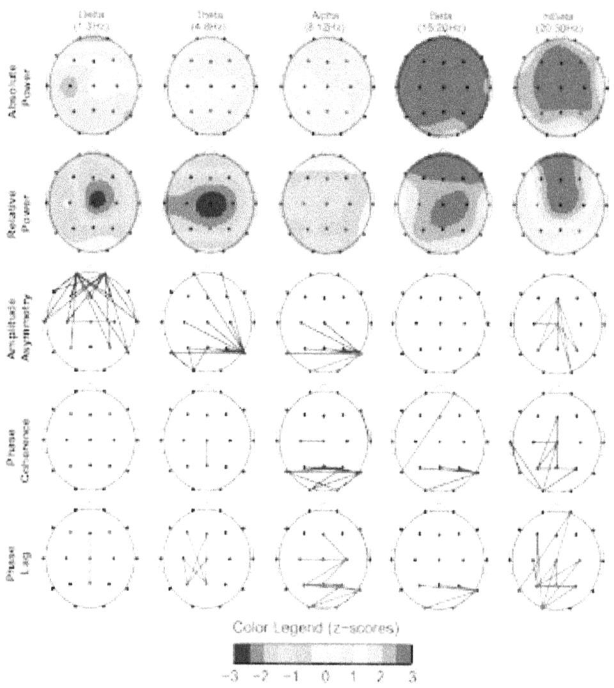

Figure 7 – Example of a BrainMap

Using qEEG-guided NFB, your brainwaves can be re-trained. Multiple training sessions will increase your brain's retention to the new way of operating. Your progress will be measured in various ways, including subsequent qEEGs and brainmaps, your self-report, and possibly other psychological questionnaires. In the following sections, we will discuss in more detail how many sessions you may need, if NFB is better than medication, and whether the results of NFB stick after training has stopped.

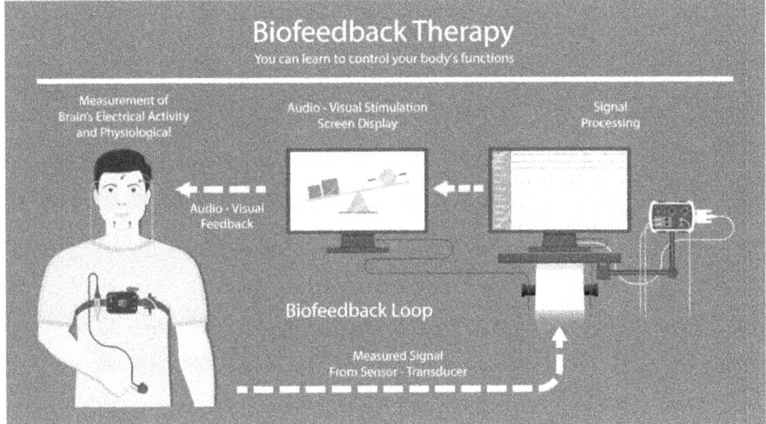

Figure 8 – Neurofeedback Loop

Neuroscientists have only recently discovered that major changes in brainwave functioning are possible. The brain's ability to reprogram itself is called neuroplasticity. This phenomenon makes the changes NFB promises possible. Researcher Orndorff-Plunkett (2017) uses the term "perturbative physiologic plasticity" to suggest that the various systems of the human body can function as an "integrated whole ... that can be *perturbed* and guided ... into different physiological states" (p.1). Orndorff-Plunkett describes the process of NFB as "self-directed neuroplasticity" that provides changes to the structure of the brain, which regulates the autonomic functioning of the brain and ultimately leads to behavioral changes.

In order to create individualized treatment protocols for neuroplasticity to make positive changes in the brain, NFB starts by observing brainwave events using electroencephalograms (EEG). Then the NFB provider devises a computer protocol based on the client's qEEG brain maps. Through a process called operant conditioning, the NFB software feeds the brain's own EEG data back to itself, conditioning brain-function toward optimization. The software uses audio and visual stimulus as the mechanism to "feed-back" the brain's patterns to itself (Smith, Collura, and Tarrant, 2014). It is typical to conduct a new qEEG every 15 to 20 sessions until the desired results are achieved. Most clients get the results they want within 40 to 60 sessions. For more severe cases, the client may need closer to 100 plus sessions. Results vary for every client.

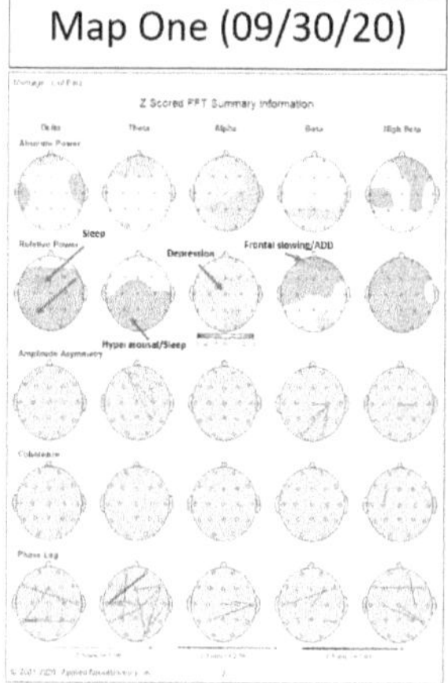

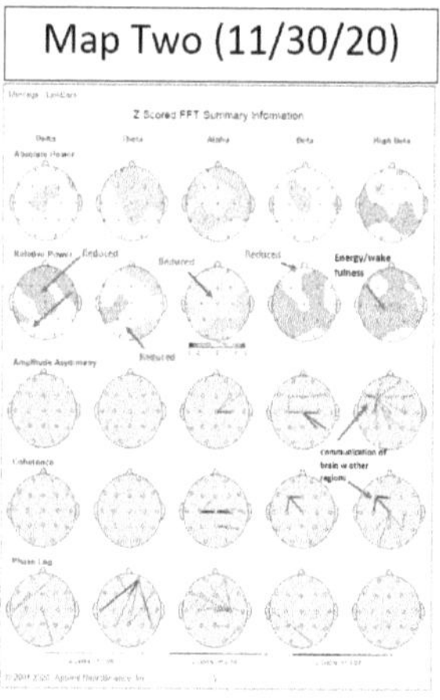

Figure 9 – qEEG Comparison

We create individualized protocols using specialized NFB software to run for about 20 sessions and target specific areas of the brain. After about 20 sessions of a protocol, another qEEG reading is taken to compare the changes in brain activity. You can see in Figure 9 two qEEGs side by side. One is pre-NFB treatment and the other is after the first 20 sessions.

As indicated by the brain maps, we see a significant decrease in slow-wave activity in the Delta and Theta bands, which help regulate sleep. At the time of Map Two, the client reported improved sleep symptoms. In the Alpha band, we see a reduction in excess energy activity in a mood regulation area of the brain, which makes sense given the client's challenges with depression. At the time of Map Two, the client reported a decrease in depressive symptoms. On the Beta band, we see an increase in fast-wave activity in the prefrontal areas, which are usually a sign of executive functioning issues, possibly ADHD. The client reported being diagnosed with ADHD as a child. At the time of Map Two, they reported a significant increase in their ability to focus and follow through with tasks.

At this point, most clients continue treatment for at least another 20-session protocol. The average total amount of sessions is 40. At the end of 40 sessions, we would expect the qEEG to show a much more regulated brain.

CHAPTER 3

NEUROFEEDBACK EFFECTIVELY TREATS ADHD

Neurofeedback has numerous endorsements and research articles that show it is an effective treatment for ADHD. However, there is no "silver bullet" solution for ADHD right now, although the research is growing. Moreover, the standard of care for children and adolescents with ADHD as recommended by the Center for Disease Control and Prevention (CDC)[3] includes parent training in behavior management, behavioral classroom interventions, and methylphenidate (e.g. ritalin) or another FDA-approved medication. For adults, as recommended by Children and Adults with Attention-Deficit/Hyperactivity Disorder (CHADD), the treatment options

---

3 https://www.cdc.gov/ncbddd/adhd/guidelines.html

suggested are medications and talk therapy. In other words, the current standards of care do not yet include NFB training.

While the CDC and CHADD have yet to endorse NFB in ADHD treatment, the most recent literature reports on the effectiveness of NFB. In some studies, the research demonstrates that NFB is more effective than the current standard of care. Arns, Clark, Trullinger, DeBeus, Mack, and Aniftos (2020) found the following:

Using a stricter version of the [American Psychological Association] guidelines, that standard neurofeedback protocols in the treatment of ADHD can be considered as well-established and 'efficacious and specific,' with medium to large effect sizes and 32–47% remission rates and sustained effects as assessed after 6–12 months. (p. 46)

In other words, of the studies they reviewed for their article, they found that nearly 50% of participants who used NFB to treat their ADHD achieved a remission of ADHD symptoms that reduced symptoms for up to 12 months.

This 2020 study was a follow up to the pronouncement that Arns and colleagues made in 2009 when they concluded that "neurofeedback treatment for ADHD can be considered 'Efficacious and Specific' (Level 5) with a large [effect size] for inattention and impulsivity and a medium ES for hyperactivity" (p.180).

Effect size (ES) is a measurement that shows clinicians how reliable a treatment is. If the ES is higher, the treatment is more reliable. ES is a way that researchers compare variables and is important when we consider claims in the research. In their type of study, a meta-analysis, the calculation of an ES includes pre- and post-treatment averages of reported changes and standard deviations from the studies included in the meta-analysis. This allows for a 95% confidence interval per study. With ES, researchers don't have to use the exact same statistical tests in each study to compare the studies to each other. Statisticians can use ES to provide a number that reflects findings reported from multiple studies with different statistics.

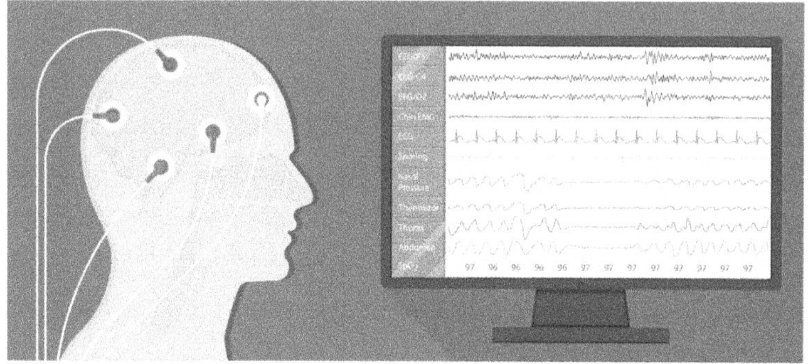

Figure 10 – Real Time qEEG

The second big finding comes from a randomized controlled trial by Duric, Assmus, Gundersen, and Elegen (2012). They found that based on reports from parents, NFB significantly improved the core symptoms of ADHD, similar to those of Ritilian, in children. They conclude that NFB is "an alternative therapy for children and adolescents with ADHD."

We provide neurofeedback to people because of this compelling research, we have seen it work, and we want to offer an effective treatment other than medication.

In numerous studies, NFB is found to work as well as medication for many people with ADHD. That is a major finding requiring further articulation, yet it provides an option for people who want to try something besides medication to manage their or their loved one's ADHD.

Lastly, there is evidence that NFB can actually change the brain such that it can normalize what we observe to be ADHD. Levesque, Beauregard, and Mensour (2006) report that neurofeedback "has the capacity to normalize the functioning of the...key neural substrate of selective attention" (p.216). Gani, Birbaumer, and Strehl (2008) support this finding. In many cases when researchers use the word "normalize" they are referring to what practitioners know is a statistical average of many people's brains. Therefore, it is normal in terms of your brain compared to a large database of brains.

Your brain will always be normal for you.

CHAPTER 4

HOW NEUROFEEDBACK WORKS FOR ADHD

Our approach to summarizing the literature is not to provide a comprehensive review. It is to provide evidence that supports using neurofeedback for ADHD to address the most common questions we get from people seeking neurofeedback, as well as from parents, spouses, partners, teachers, administrators, and professional athletes. Some will find us biased. We are providers of neurofeedback who specialize in ADHD, after all. But we address some of the well-known criticisms so that those who are eager to do a more thorough literature review will have a head start in doing so.

How many sessions will I need before neurofeedback works?

That's the million-dollar question. The truth is that no one will ever know how many sessions any one person needs. Both of us have trained people for as few as five sessions and as many as 100 sessions or more. When we convey this to clients and colleagues, we see flashes of panic in their eyes because they are doing the math for cost, time, and disruption to their schedules. For our clients who decide to move forward with the treatment, nothing outweighs the relief they get from neurofeedback training. A bit later we will provide a more in-depth understanding of the cost comparison of NFB treatment versus standard of care treatment for ADHD.

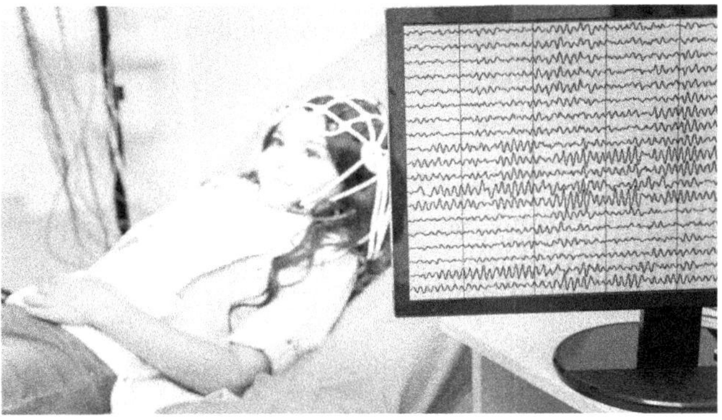

Figure 11 – Child During qEEG Collection

Diving deeper into the literature, Fritson, Wadkins, Gerdes, and Hof (2007) reported that "response control may improve in as few as 20 EEG biofeedback sessions" (p.1). Lastly, in this brief summary, Mayer, Wyckoff, Schulz, & Strehl's (2012) article offers "a promising outlook to the outcome after the completion of 30 sessions" (p.37).

Do the effects of neurofeedback last overtime?

For the most part, our clients report that the changes stick.

While some come in for "tune-ups" from time to time, typically, after the initial treatment is completed, the gains that clients report remain. There is less literature about this, as the follow up studies are expensive and logistically challenging to conduct. However, there is good evidence supporting the notion that the changes last over time. Monastra, Monastra, and George (2002) report that gains from neurofeedback were retained three years after the completion of the neurofeedback training. Gani, Birbaumer, and Strehl (2008) report that the clinical effects of neurofeedback training were stable after six months. They also reported that two years after the initial study was completed EEG self-regulation skills were retained and all improvements in behavior and attention were retained with "significant reductions in the number of reported problems and significant improvement in attention" (p.209).

In a more recent meta-analysis, Arns, Clark, Trullinger, DeBeus, Mack & Aniftos (2020) reported that "remission rates seen across studies between 32 and 68% for controlled treatments, reflect clinically meaningful remission rates" (np). More specifically, they report "remission rates for neurofeedback (32–47%) in this context are reassuring, and there is the clear finding that effects of neurofeedback are sustained without further treatment" (np). It is exciting to see how reduced ADHD symptoms stick with NFB.

Medication and neurofeedback

Often people come for neurofeedback training after they have tried other methods. We regularly see clients who are taking medications. Medication does have an impact on NFB in most cases. See Table 1 below for a summary of how certain medications can impact brainwave activity.

Table 1: The influence of medications on brain waves.

Drug	Increases	Decreases
Adderall	Beta between 12-26 Hz	Delta and Theta
Ativan	Beta between 20-30 Hz	Alpha
Haldol	Delta, Theta, and Beta above 20 Hz	Alpha and Beta below 20 Hz
Lithium	Theta and mild increase in Beta	Mild decrease in Alpha
Opiates	High amplitude slow Al-pha in 8 Hz	None reported
Prozac	Mild increases in Beta at 18-25 Hz	Frontal Alpha
Seroquel	Global Beta	Alpha and Beta below 20 Hz
THC	Frontal Alpha	Beta and high Beta
Vyvanse	Beta between 12-26 Hz	Delta and Theta
Xanax	Beta between 20-30 Hz	Alpha
Zoloft	Mild increases in Beta at 18-25 Hz	Frontal Alpha

Adapted from Stress Therapy Solutions (2019)

There are two studies included in this survey of the literature on the influence of neurofeedback and medication. Monastra (2005) reported that 80% of the clients whose treatment included EEG biofeedback were able to decrease daily dosage of stimulant by at least 50%… .by contrast, none of the clients who did not receive EEG biofeedback was able to reduce dosage [and that] 85% increased [their] dose. (p.69)

Fuchs, Birbaumer, Lutzenberger, Gruzelier, and Kaiser (2003) conducted a study in which they had some children undergo neurofeedback training and some children take methylphenidate (e.g. Ritalin). While they found that both the children who had neurofeedback training and those that took medicine had improvements on their continuous performance tests, they concluded that "neurofeedback was efficient in improving some of the behavioral concomitants of ADHD in children whose parents favored a nonpharmacological treatment" (p.1). **While this finding might seem underwhelming at first, it suggests that neurofeedback could be as effective as medication in children.** More research is needed to confirm this finding, but it is promising and exciting for those who want to reduce or stop their medications.

As with all research, we should be cautious to take the conclusions too far. Moreover, the translation of research into practice is quite complex given that clinics aren't lab settings, the demographic and clinical profile of each person we treat may not be represented in the literature, and each individual has unique experiences that aren't measured or reported in the classical research.

Children and ADHD

An overwhelming majority of the literature on NFB and ADHD was conducted with children and people up to age 19. There are also a growing number of articles showing the positive results of neurofeedback training on adults with ADHD (Valdez, 1985; Mayer, Wyckoff, Schulz, & Strehl, 2012).

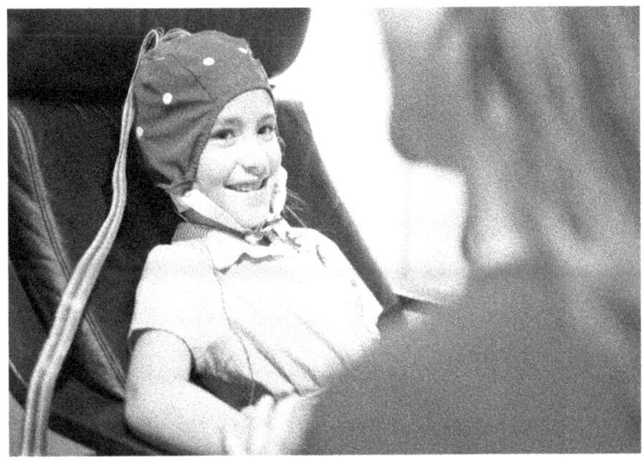

Figure 12 – Child During Neurofeedback Training

Pakdaman, Irani, Tajikzadeh, and Jabalkandi (2018) conducted a smaller study and reported that NFB can help children whose parents prefer not to use Ritalin. More importantly, they suggest that Ritalin and neurofeedback in combination are more efficient. We will cover the medication effects on NFB in more detail later on.

In a study by Monastra, Monastra, and George (2002), 100 children participated in a study that provided medication, parent counseling, academic support, and NFB to 51 of the children while 49 received everything but the NFB. **The results showed that only the children who received NFB, and didn't use Ritalin, retained the progress they made using NFB.** In the follow up study to this initial article, Monastra (2005) reported that after three years only the children who received NFB sustained gains on all measures while unmedicated. **Furthermore, 80% of the NFB group had reduced their medications by 50% or more.** And almost as striking, children that didn't get NFB didn't reduce their dosage of stimulant medication—85% had increased their dosage.

Carmondy, Radvanski, Wadhwani, Sabo, and Vergara (2001) reported on a smaller sample of 16 children who demonstrated improved scores on the continuous performance tests for children with ADHD, significantly reduced **hyperactivity and impulsivity, and significantly improved ratings by their teachers.**

Self-Reported Measures

Numerous studies showed improvements in self-reported measures in children and adults (Heinrich, Gevensleben, Freisleder, Moll, & Rothenberfer, 2004; Rossiter, 2004; Bakhshayesh, Hansch, Wyschkon, Rezai, & Esser, 2011; Mayer, Wyckoff, Schulz & Strehl, 2012; Gevensleben, Kleemeyer, Rothenberger, Studer, Flaig-Röhr, et al., 2014). Self-report measures are typically questionnaires or clinician-guided tests that are administered to the client. These include measures that provide reliable assessment of a person's ADHD symptoms and often include subscales that reflect executive functioning in people with ADHD. While there is some dispute about how unbiased these self-report measures are, they are frequently used in mental health research and treatment and are frequently used to establish a baseline of symptoms while using NFB for ADHD.

Continuous Performance Tests (CPT)

Likewise, there was representation in the literature that CPTs showed improvements for people with ADHD who were treated with NFB (Carmondy, Radvanski, Wadhwani, Sabo, & Vergara, 2001; Monastra, Monastra, & George, 2002; Fuchs, Birbaumer, Lutzenberger, Gruzelier, & Kaiser, 2003; Rossiter, 2004). A CPT is a type of neuropsychological test that measures things like auditory and visual response control, auditory and visual attention, and sustained attention. They are computer-based tests where the client is given instructions, usually to click or not click a mouse or press the spacebar on the keyboard, when a specific cue is given or not given. These usually take about 20 minutes to complete and are typically reliable measurement to see change in symptoms in people who use NFB. These are completed by the person receiving NFB. They can also include measures of vigilance, acuity and elasticity, focus, dependability, sustainability, speed, swiftness, prudence, reliability, consistency, stamina, fine motor activity, comprehension, steadiness, reliability, and persistence.

Parental/Guardian Ratings

When working with children with ADHD, their parents or guardians provide vital information about the child's mood, behavior, and general wellbeing. A number of studies show improved parental ratings in children with ADHD after NFB treatment. Cortese et al. (2016) found that both parent and teacher rating scales confirmed a significant benefit from NFB. Moreover, Bakhshayesh, Hansch, Wyschkon, Rezai, and Esser (2011) found that parents shared significant decreases in ADHD symptoms and improvements in attention with the children who received NFB versus those who didn't get NFB. Furthermore, Drechsler et al. (2007) found that the children in the neurofeedback treatment group in their study had better parental ratings, particularly regarding attention and cognition-related domains, than children who were treated with Cognitive Behavioral Therapy. Leins et al. (2007) also found that parents reported "significant behavioral and cognitive improvements" among children in their study. In their randomized control trial,

Gevensleben et al. (2009) reported that parents rated improvements in the following: overall ADHD symptoms; inattention; hyperactivity/impulsivity; oppositional behavior; and delinquent and physical aggression among children treated with ADHD as compared to a different intervention called "Skillies." That is a lot of evidence to suggest that parents reported improvements in their children who received NFB training.

Teacher's Reports

Teacher's reports on improvements are also an important part of realizing change. In other words, third parties who know the client well can provide valuable feedback related to the impact of NFB. While we would like to rely only on what clients tell us, it is important to include their parents and third parties whenever possible. The literature reports a decent amount of articles that show an improved teacher rating in children being treated for ADHD with neurofeedback (Carmondy, Radvanski, Wadhwani, Sabo, and Vergara, 2001; Strehl, Leins, Goth, Klinger, Hinterberger, and Birhaumer, 2006; Drechsler, Straub, Doehnert, Heinrich, Steinhausen, and Brandeis, 2007; Leins, Goth, Hinterberger, Klinger, Rumpf, and Strehl, 2007; Gevensleben, Holl, Albrecht, Vogel, Schlamp, Kratz, Studer, Rothenberger, Moll, and Heinrich, 2009; Wangler, Gevensleben, Albrecht, Studer, Rothenberger, Moll, and Heinrich 2011; Cortese, Ferrin, Brandeis, Holtmann, Aggensteiner, Daley, et al., 2016.) This is important as it lends credence to the observations of change from someone other than the client and/or the parents/guardians.

Women and ADHD

Figure 13 – Eyes Closed Neurofeedback Training

We fully intended to present evidence on the differences between men and women who have ADHD and want to use NFB for treatment. We were hopeful that there would be good evidence and we were skeptical that there would be any evidence on genders outside of men and women, because, broadly speaking, science isn't keeping up with the multitude of ways people are empowered to identify. But we were severely let down in our attempts to find anything to help you understand how NFB treatment differs for women versus men.

All the research we reported on above gives the impression that men, women, and all genders are included equally. But, sadly, that is not the case. In our review of the literature, we found only one article that was specific about ADHD and NFB treatment in women and no literature that used a large list of gender identities. The only relevant citation we found confirmed what we hoped would not be true. Dupuy and Clarke (2012) explain "the lack of comprehensive research draws attention to the necessity for sex-specific EEG research within ADHD populations" (pg.49).This is frustrating as providers who seek to treat all genders equally. Moreover, it is frustrating because it is incredibly rewarding to provide each individual with unique care. Functionally, applying what is in the research to each individual's specific case is called translational research, where we translate the research literature and apply it to individual cases. NFB is a powerful tool to do this because each brain is unique even if there is literature that includes specific aspects of an individual case. Without scientific study of gender specific responses to NFB, we can't rely on the broader scientific community to help as much as they should.

In practice, this means that our anecdotal experience is the most reliable measure in treatment across the genders. We treat all people and have had positive outcomes with many. And, while it is outside the scope of this book to detail the recent advances in understanding the differences between men and women who are diagnosed with ADHD, we want to at least point you in the right direction.

There is a growing body of literature that seeks to detail the differences between men and women with ADHD. It is important to note that ADHD in men, women, girls, and boys presents differently. Providers need to be aware of this and people seeking clarity regarding their symptoms or the symptoms of children in their care need to know there is a difference. Author Kathleen Nadeau is an expert in this and has written books such as *Understanding Women with AD/HD, Understanding Girls with ADHD: How they feel and why they do what they do,* and *When Moms and Kids Have ADD.*

The main point to take away regarding ADHD and women is that the condition is misunderstood and under-diagnosed. Many women come to us for ADHD testing who have had ADHD symptoms all their lives but were told by family and authority figures that they were exaggerating or making up excuses for being lazy. Or worse, perhaps, girls with ADHD Inattentive Type fall through the cracks of mental health diagnosis and treatment because they are so quiet, off in their own dreamworld. Because they aren't acting out like most boys with ADHD, many girls are often undiagnosed until later in adulthood.

Critics of Neurofeedback

There are critics of NFB treatment for ADHD. The most prominent article is by Pigott and Cannon (2014). It covers their main concerns about supporting NFB as an effective treatment for ADHD. They lay out important considerations when trying to decipher the research about ADHD and NFB, including that better evidence, more rigorous study designs, and comparisons to other methods for treatment for ADHD are needed. They also take issue with the measurements that are used to report change, including self-report, parental-report, and teacher-report measures. They suggest there might be false positives, false negatives, and biases that are not accounted for in the study designs.

As we can see, most of the criticism of NFB comes from arguments we agree with. NFB is still expensive (we will cover this in more detail in Chapter 5) and should be covered by insurance companies. NFB is also a time-consuming process, taking up to 100 or more sessions, which can cause treatment to last several months or more. However, compared to medication, which is often considered a life-long treatment, NFB is a comparatively shorter process and often cheaper over time than the National Institutes of Mental Health (NIMH) recommended treatments for ADHD.

CHAPTER 5

NEUROFEEDBACK IS CHEAPER THAN THE STANDARD OF CARE FOR ADHD

What is the typical cost of NFB training? We get this question a lot because NFB seems like a more expensive process to the average person than simply taking one medication daily. But when you look deeper, a different narrative emerges. **NFB treatment is actually less expensive than the CDC and the National Institute of Mental Health (NIMH) recommended standard of care treatment for ADHD with medication and therapy.**

Zhao et al. (2019) report that the average cost to a family with a child who requires ADHD treatment is $15,036 whereas a child without ADHD costs a family $2,848. In other words, a child with ADHD may cost a family 80% more than a child without ADHD. Furthermore, they reported that this was not just for ADHD alone but that the significant cost savings remained for children when they controlled for intellectual functioning, oppositional defiant symptoms, or conduct problems.

As reported by Pigott, De Biase, Bodenhamer-Davis, and Davis (2013), the seminal NIMH study[4] recommends treatment for ADHD that averages around $11,000. This is consistent with what Sturm et al. (2001) estimated $11,681 cost for children between one and 17 years old. Moreover, Pigott, De Biase, Bodenhamer-Davis, & Davis (2013) report the follow-up NIMH study found that 6 years after intensive parent training, and optimal medication treatment, there was still moderate-to-severe symptoms severity and impairment with 89% of the children first treated in preschool. (p. 9)

This is an important finding and suggests that alternative and better interventions for ADHD are desperately needed. It also makes the information about the sustained effect of NFB training in reducing symptom severity for people with ADHD even more pertinent.

You might be thinking, *if my kid or I have ADHD this is moot since we have to get treatment.* And you'd be on the right track. Not all people with ADHD get treatment and the costs of not getting mental health care are not well documented. But we can tell you that we see so many adults who report their parents "didn't believe" in ADHD or didn't want to get them tested or

---

4 https://www.nimh.nih.gov/funding/clinical-research/practical/mta/the-multimodal-treatment-of-attention-deficit-hyperactivity-disorder-study-mta-questions-and-answers.shtml

give them medications or receive any form of treatment. We have seen first-hand how this translates into lower grades, lower self-esteem, more checkered job history, fewer promotions, more mental health care, more trips to the doctor, more self-help books, personal growth retreats, etc..

Looking closely at the different costs of treatment for ADHD, Arns et al. (2020) estimate that 30 to 40 sessions of NFB costs between $4,000 and $6,000. The variance is likely due to geographical price differences and might be covered in full or in part by your insurance company. This also doesn't include any reference to a pre and post qEEG which can add more cost to the treatment.

Arns, Clark, Trullinger, DeBeus, and Mack (2020) go on to say if a pill for ADHD medication is about two dollars per day the overall cost of medication for a five-to-10-year period would be between $3,500 and $7,000. We aren't sure why they stopped their estimate at 10 years as we have clients who have taken medications for much longer.

Now, if you still think that medication is cheaper, consider this. What many of us forget in our cost calculations are ongoing medical appointments, cost of treating side effects, cost of parental training, behavioral counseling, ADHD coaching, summer camps, etc.. While they don't offer a final tally of the associated costs, suffice it to say that on the face of things, NFB is probably going to be much cheaper than the NIMH recommended treatments. If you combine this with the almost 50% remission rates in ADHD symptoms with NFB training versus the NIMH's report that six years post treatment 89% of children in the study still had "moderate-to-severe symptoms severity and impairment," we think the comparison comes into clearer focus with **the clear winner being NFB training.** Furthermore, as referenced above, Monastra (2005) reported in a follow up study, that 80% of people in the study who received NFB for ADHD decreased their daily dose by at least 50% whereas, 85% of those who did not use NFB increased their dose. The cost of NFB might be more upfront but will pay off in the long run. **In other words, NFB for ADHD is an investment in the future of your brain and your savings account.**

Remote NFB

Remote NFB is a newer development that has provided a reliable training option for people who prefer not to come into our offices when they train. This was more widely utilized during the COVID-19 pandemic. This provides more flexibility for clients, cuts down on commute time, and allows adults and children to receive NFB training in the comfort of their own homes. Additionally, it allows for more than one member of the household or office to receive tailored NFB treatment without having to travel to your NFB provider's office. Instead of waiting for their children to emerge from their NFB session, parents could be doing something at home while their child trains.

Our advanced methods of NFB are "prescription strength," the best treatment available. Remote NFB offers exactly the same level of care as coming into our offices. But some remote NFB is "over the counter." Therefore, it is important to be discerning about remote NFB training because there are some systems out there that are cheap and you can use on your own, with an app on your smartphone, and even some YouTube videos that claim to offer "brain training." We like to characterize these systems as "over the counter" NFB, whereas we provide "prescription strength" NFB. Moreover, some of these might be guided by a NFB provider who may use platforms that do not protect your health information and thereby make even your brainwave information available for people to see or even sell to marketers. That is not something we recommend given the sensitivity of your healthcare information and that of your brainwaves. We strongly suggest using NFB administered by a trained NFB provider.

When you work with a trained and licensed NFB provider you can be sure you're getting prescription strength NFB. While there are many ways to set up a NFB station in your own home or office, the easiest way to do it is with a reliable internet connection, then the NFB provider supplies the rest. Of course, you will need a computer, monitors, cables, neurofeedback hardware, software, etc., but that can all be provided to you by the NFB provider. Alternatively, you can provide some of the components of the home training system if you already have them at home or at your office.

In remote NFB, the client typically rents the equipment for a monthly rental fee. Additionally, the NFB provider is likely to still charge an appointment fee and in some cases ask you to put down a deposit on the equipment. To ensure that you are covered with what can amount to $10,000 or more of equipment, we suggest you notify your homeowners or renters insurance providers and list the equipment as a part of your insurance plan. You don't want to end up paying for the full cost of the equipment if you damage it. It is easier than you think to spill your morning coffee or a glass of water on the equipment. If the NFB provider asks you to put down a deposit and there is a rental fee, the rental fee is typically deducted from the deposit so that you get your deposit back (minus your rental fee) when your treatment is over. For example, if the rental fee is $500 per month, the deposit is $2,500, and you rent the equipment for three months, you would get back $1,500 (3 months * 500) or $1,000. In some cases, clients have simply purchased their own set up because either they can afford it and/or because they find it easier.

Once your home or office system is set up, the NFB provider can remote control into the computer you're using for your remote training and run the software to administer the training. The process is very user-friendly. The only mildly challenging part is learning how to set up the NFB Electro-Cap. While this seems like it takes a lot of practice and patience, it is something you can learn quickly from your NFB provider, through a training video and meeting with your NFB provider. Clients who take pictures or a video of themselves during setup are more likely to remember how to set themselves up next time. We are able to see the quality of the connection of the electrode on your scalp in real time and help you adjust them as necessary. This is something we wouldn't trust with "over the counter" remote training solutions.

CHAPTER 6

WHAT YOU CAN EXPECT DURING NEUROFEEDBACK TRAINING SESSIONS

Clients new to NFB tend to have worries because they don't know what to expect. That's a normal human reaction to something new. Our goal in this chapter is to walk you through what a typical neurofeedback session with us would look like. That way, if you or your child come to us for training, you'll be much more prepared and at ease.

First of all, NFB is a noninvasive treatment. Nothing is ever shot into your brain by electric pulses, magnets, or in any other form or manner. In fact, some clients voice their concern that they don't have to do anything during the training session. So, sit back, relax, and enjoy your brain healing itself by learning new patterns through neurofeedback.

When you arrive at our offices, you'll be greeted by a NFB provider. Going to the bathroom is a great idea before your session begins so that you don't end up walking down the hall with a bunch of electrodes on your head. Take a nice sip of your preferred beverage and remove your phone and any other electronic devices from your pockets. Of course, if you have medically required devices, don't remove them. But let your NFB provider know.

Next, your NFB provider will review the plan for the session and answer any questions you may have. During this portion of the appointment, it is important to let your provider know of any observations you have, or others have about you since your last visit. This information is extremely helpful to continue to tailor the training to you.

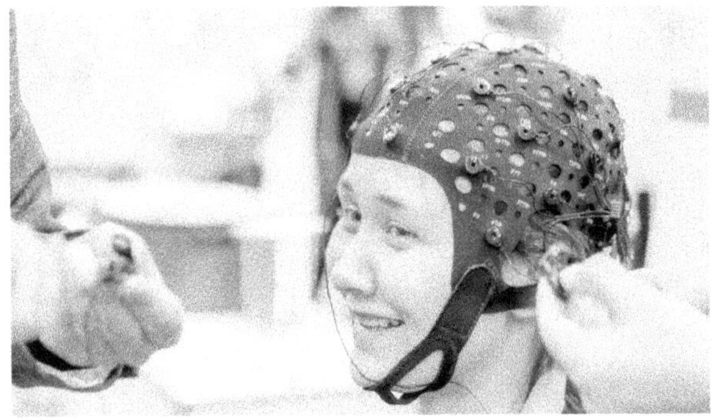

Figure 14 – Preparing the Electrodes

The NFB provider will then prepare your scalp for the electrodes by using a paste, gel, or possibly using a newer dry electrode cap. You should know that your hair might get a bit messy. It's typical for us to wipe off the gel using a baby wipe. Seasoned clients frequently bring a head covering to wear on the way out of the office. In some cases, you might want to make a stop at the local hairdresser for a wash and blow dry or just push through those double glass doors on to the street with all the confidence and messy hair you can muster.

Once your scalp is prepared and the electrodes are positioned, the training will begin. If you feel any pain or discomfort, please let your NFB provider know as there are ways to ensure your comfort without affecting the electrodes. Once the electrodes or the Electro-Cap are placed, you can sit back and relax.

The training might take various forms depending on the type of training you're receiving. You might play a video game, sit with your eyes open or closed, or watch a movie or show. You'll almost always hear a sound in the background, and you might also see a dimming or brightening effect on a screen if you're watching something. This is the feedback, and you don't have to "do anything," per se, to make it come or go. Your brain will do the work.

CHAPTER 7

WHY YOU SHOULD TRY NEUROFEEDBACK FOR ADHD

We hope that we have shown you how effective, affordable, and accepted neurofeedback is for treating ADHD. We want you to feel empowered to make an informed decision about your own or your family's mental health care. If you don't have this information or the information is written in such a way that is too hard to understand, most people are likely to just go along to get along. But living with untreated ADHD and mental health issues can be devastating. Instead, having the knowledge to make an informed decision will make us better consumers of mental health information and care.

We recommend NFB because it is supported by the literature. As you can see, there is plenty of literature that supports the use of neurofeedback for ADHD. The International Society for Neuroregulation and Research (ISNR) has a very long bibliography on the uses of neurofeedback and mental health conditions. You can find that here: https://isnr.org/isnr-comprehensive-bibliography.

Second, we recommend neurofeedback because it is an objective measure of how your brain is functioning and how your symptoms are presenting. Objectivity allows for a cleaner measurement of your progress and improvement. Coupled with other forms of measurements, even if they are self-reported, increases the likelihood that your intervention is tailored to you. In the past, the title of our speaking engagements about NFB has been "See the change you wish to be" after the famous Gandhi quote, "Be the change you wish to see." With qEEG-guided NFB you can actually see yourself becoming who you want to be.

Third, we recommend NFB because the cost is less expensive than the alternatives in the long run. The costs of training can have an impact, but the long-term effects stick and medications can be reduced or stopped completely. We all know that medications have well-documented side effects that can be upsetting to many. With NFB we can move you closer than you've ever been to having no medications, doing better at work, and being able to spend whatever extra cash you have on yourself or the people you love.

Fourth, NFB can treat not only ADHD but a wide variety of mental health and medical conditions. In our practices it is common for clients to tell us that their blood pressure normalized, and they are now off their blood pressure medication, that their pain subsided, that they are sleeping better, that they don't get angry as fast, and that they love themselves and their loved ones more deeply.

Figure 15 – Explaining qEEG Results

Lastly, we recommend NFB for those drawn to cutting-edge mental health interventions. Very soon NFB will be as common a health treatment option as medication. While it has been around for half a century, it is still considered "new." Therefore, there is an appeal to doing the newest thing. How many times have you had that person say, "I know this is a little out of the box, but I tried _________ and it really worked for me." Either you've heard someone recommend NFB or you want to be that person who tries it out so you can share your experience with others. Being a leader in your community is a great gift.

REFERENCES

1. American Psychiatric Association. (2013). *Diagnostic and statistical manual of mental disorders* (5th ed.). Washington, DC: Publisher.
2. Arns, M., Clark, C.R., Trullinger, M., deBeus, R., Mack, M., Aniftos, M. (2020). Neurofeedback and Attention-Deficit/Hyperactivity-Disorder (ADHD) in Children: Rating the Evidence and Proposed Guidelines. Appl Psychophysiol Biofeedback. Jun;45(2):39-48. doi: 10.1007/s10484-020-09455-2.
3. Arns, M., de Ridder, S., Strehl, U., Breteler, M., Coenen, A. (2009). Efficacy of neurofeedback treatment in ADHD: the effects on inattention, impulsivity and hyperactivity: a meta-analysis. Clin EEG Neurosci. Jul;40(3):180-9. doi: 10.1177/155005940904000311.
4. Arns, M. Gunkelman, J., Breteler, M, & Spronk, D. (2008). EEG phenotypes predict treatment outcomes to simulants in children with ADHD. *Journal of Integrative Neuroscience, 07(*03), 421-438. Doi: 10.1142/s0219635208001897.
5. Arns, M., Heinrich, H., & Strehl, U. (2014). Evaluation of neurofeedback in ADHD: The long and winding road. *Biological Psychology, (95)*, 108-115.
6. Bakhshayesh, A.R., Hansch, S., Wyschkon, A., Rezai, M.J. & Esser, G. (2011). Neurofeedback in ADHD: a single-blind randomized controlled trial. *European Child & Adolescent Psychiatry, 20*, 481-491.
7. Biofeedback Certification International Alliance.

https://www.bcia.org/i4a/pages/index.cfm?pageid=3524. Accessed 11/09/2020 at 1200.

8. Carmondy, D.P., Radvanski, D.C., Wadhwani, S., Sabo, J.J., Vergara, L. (2001). EEG biofeedback training and attention-deficit/hyperactivity disorder in an elementary school setting. Journal of Neurotherapy, 4, 5-27.
9. Cortese, S., Ferrin, M., Brandeis, D., Holtmann, M., Aggensteiner, P., Daley, D., et al. (2016). Neurofeedback for attention-deficit/hyperactivity disorder: Meta-Analysis of clinical and neuropsychological outcomes from randomized controlled trials. *Journal of the American Academy of Child & Adolescent Psychiatry,55*(6), 444–455.
10. Crichton, A. (1798). An inquiry into the nature and origin of mental derangement: comprehending a concise system of the physiology and pathology of the human mind and a history of the passions and their effects. Cadell T Jr, Davies W, London [Reprint: Crichton, A. (2008) An inquiry into the nature and origin of mental derangement. On attention and its diseases. J Atten Disord 12:200–204]
11. Dupuy, E. F., & Clarke, A. (2012). EEG activity in females with attention-deficit/ hyperactivity disorder.
12. Duric, N.S., Assmus, J., Gundersen, D., Elgen, I.B. (2012) Neurofeedback for the treatment of children and adolescents with ADHD: a randomized and controlled clinical trial using parental reports. BMC Psychiatry. Aug 10;12:107. doi: 10.1186/1471-244X-12-107. PMID: 22877086; PMCID: PMC3441233.
13. Drechsler, R., Straub, M., Doehnert, M., Heinrich, H.,

Steinhausen, H., & Brandeis, D. C. (2007). Controlled evaluation of a neurofeedback training of slow cortical potentials in children with attention deficit/hyperactivity disorder. Behavioral and Brain Functions, 3, 35.

14. Fuchs, T., Birbaumer, N., Lutzenberger, W., Gruzelier, J.H., & Kaiser, J. (2003). Neurofeedback treatment for attention-deficit/hyperactivity disorder in children: A comparison with methylphenidate. Applied Psychophysiology and Biofeedback, 28(1), 1-12.
15. Fritson, K. K., Wadkins, T. A., Gerdes, P., & Hof, D. (2007). The impact of neurotherapy on college students' cognitive abilities and emotions. Journal of Neurotherapy, 11(4), 1–9.
16. Gani, C., Birbaumer, N., Strehl, U. (2008). Long term effects after feedback of slow cortical potentials and of theta-beta-amplitudes in children with attention-deficit/hyperactivity disorder (ADHD). International Journal of Bioelectromagnetism, 10(4) 209-232
17. Gevensleben, H., Holl, B., Albrecht, B., Vogel, C., Schlamp, D., Kratz, O., Studer, P., Rothenberger, A., Moll, G.H., Heinrich, H. (2009). Is neurofeedback an efficacious treatment for ADHD? A randomised controlled clinical trial. J Child Psychol Psychiatry. 2009 Jul;50(7):780-9. doi: 10.1111/j.1469-7610.2008.02033.x. Epub 2009a Jan 12. PMID: 19207632.
18. Gevensleben, H., Holl, B., Albrecht, B., et al. (2009b). Distinct EEG effects related to neurofeedback training in children with ADHD: A randomized controlled trial. International Journal of Psychophysiology, 74, 149-157.
19. Gevensleben, H., Kleemeyer, M., Rothenberger, L.G.,

Studer, P., Flaig-Röhr, A., Moll, G.H., Rothenberger, A., Heinrich, H. (2014). Neurofeedback in ADHD: further pieces of the puzzle. Brain Topogr.Jan;27(1):20-32. doi: 10.1007/s10548-013-0285-y.

20. Heinrich, H., Gevensleben, H., Freisleder, F.J., Moll, G.H., Rothenberger, A. (2004). Training of slow cortical potentials in attention-deficit/hyperactivity disorder: evidence for positive behavioral and neurophysiological effects. Biol Psychiatry. Apr 1;55(7):772-5. doi: 10.1016/j.biopsych.2003.11.013. PMID: 15039008.
21. International Society for Neurofeedback & Research (2009). https://isnr.org/what-is-neurofeedback. Accessed 11/09/2020 at 1208.
22. Johnstone, J., Gunkelman, J. & Lunt, J. (2005). Clinical database development: characterization of EEG phenotypes. *Clinical EEG and Neuroscience, 36*(2), 99-107. Doi: 10.1177/155005940503600209.
23. Leins, U., Goth, G., Hinterberger, T., Klinger, C., Rumpf, N., Strehl, U. (2007). Neurofeedback for children with ADHD: a comparison of SCP and Theta/Beta protocols. Appl Psychophysiol Biofeedback. Jun;32(2):73-88. doi: 10.1007/s10484-007-9031-0. Epub 2007 Mar 14. PMID: 17356905.
24. Levesque, J., Beauregard, M., & Mensour, B. (2006). Effect of neurofeedback training on the neural substrates of selective attention in children with attention-deficit/hyperactivity disorder: A functional magnetic resonance imaging study. Neuroscience Letters, 394, 216-221.
25. Lubar, J.F., Swartwood, M.O., Swartwood, J.N. et al.

(1995). Evaluation of the effectiveness of EEG neurofeedback training for ADHD in a clinical setting as measured by changes in T.O.V.A. scores, behavioral ratings, and WISC-R performance. *Biofeedback and Self-Regulation* 20, 83–99. https://doi.org/10.1007/BF01712768

26. Marzbani, H., Marateb, H. R., & Mansourian, M. (2016). Neurofeedback: A Comprehensive Review on System Design, Methodology and Clinical Applications. *Basic and clinical neuroscience*, *7*(2), 143–158. https://doi.org/10.15412/J.BCN.03070208
27. Mayer, K, Wyckoff, S.N., Schulz, U., & Strehl, U. (2012). Neurofeedback for Adult Attention-Deficit/Hyperactivity Disorder: Investigation of Slow Cortical Potential Neurofeedback—Preliminary Results, Journal of Neurotherapy: Investigations in Neuromodulation, Neurofeedback and Applied Neuroscience, 16:1, 37-45, DOI: 10.1080/10874208.2012.650113
28. Monastra VJ. (2005). Electroencephalographic biofeedback (neurotherapy) as a treatment for attention deficit hyperactivity disorder: rationale and empirical foundation. Child Adolesc Psychiatr Clin N Am. 2005 Jan;14(1):55-82, vi. doi: 10.1016/j.chc.2004.07.004. PMID: 15564052.
29. Monastra, V.J., Monastra, D.M. & George, S. (2002). The effects of stimulant therapy, EEG biofeedback, and parenting style on the primary symptoms of attention-deficit/hyperactivity disorder. Applied Psychophysiology and Biofeedback, 27, 231-249.
30. Orndorff-Plunkett, F., Singh, F., Aragón, O. R., & Pineda, J. A. (2017). Assessing the Effectiveness of

Neurofeedback Training in the Context of Clinical and Social Neuroscience. *Brain sciences*, 7(8), 95. https://doi.org/10.3390/brainsci7080095

31. Pakdaman, F., Irani, F., Tajikzadeh, F. et al. *(2018).* The efficacy of Ritalin in ADHD children under neurofeedback training. *Neurol Sci* 39, 2071–2078. https://doi.org/10.1007/s10072-018-3539-3
32. Pigott, E. H. & Cannon, R. (2014). Neurofeedback Requires Better Evidence of Efficacy Before It Should Be Considered a Legitimate Treatment for ADHD: What is the Evidence for this Claim?. Journal of NeuroRegulation, 1(1), 25-45.
33. Pigott, H.E., De Biase, L., Bodenhamer-Davis, E., Davis, R.E. (2013). The Evidence-Base for Neurofeedback as a Reimbursable Healthcare Service to Treat Attention Deficit/Hyperactivity Disorder. International Society of Neurofeedback and Research.
34. Rossiter T. (2004). The effectiveness of neurofeedback and stimulant drugs in treating AD/HD: Part II. Replication. Appl Psychophysiol Biofeedback, 29(4) 233-243.
35. Strehl, U., Leins, U., Goth, G., Klinger, C., Hinterberger, T., & Birhaumer, N. (2006). Self-regulation of slow cortical potentials: A new treatment for children with attention-deficit/hyperactivity disorder. Pediatrics, 118, 1530-1540.
36. Sturm, R., Jeanne S.R., Yuhua B., Bradley D.S, Kanika, K., Winnie, Z., & Feng, Z. (2001). Mental Health Care for Youth: Who Gets It? How Much Does It Cost? Who Pays? Where Does the Money Go?. Santa Monica, CA: RAND Corporation, 2001.

https://www.rand.org/pubs/research_briefs/RB4541.html.

37. Studer, P., Kratz, O., Gevensleben, H., Rothenberger, A., Moll, G. H., Hautzinger, M., & Heinrich, H. (2014). Slow cortical potential and theta/beta neurofeedback training in adults: effects on attentional processes and motor system excitability. *Frontiers in human neuroscience*, *8*, 555. https://doi.org/10.3389/fnhum.2014.00555
38. Adapted from Stress Therapy Solutions. (2019). Neurofeedback Bootcamp for Beginners: Psychopharmacological considerations. PowerPoint Presentation.
39. Wangler, S., Gevensleben, H., Albrecht, B., Studer, P., Rothenberger, A., Moll, G.H., & Heinrich, H. (2011). Neurofeedback in children with ADHD: Specific event related potential findings of a randomized controlled trial. Clinical Neurophysiology, 122, 942-950.
40. Valdez, M. (1985). Effects of biofeedback-assisted attention training in a college population. Biofeedback & Self-Regulation, 10(4), 315–324.
41. Zhao, X., Page, T. F., Altszuler, A. R., Pelham, W. E., 3rd, Kipp, H., Gnagy, E. M., Coxe, S., Schatz, N. K., Merrill, B. M., Macphee, F. L., & Pelham, W. E., Jr. (2019). Family Burden of Raising a Child with ADHD. *Journal of abnormal child psychology*, *47*(8), 1327–1338. https://doi.org/10.1007/s10802-019-00518-5.

ABOUT THE AUTHORS

Schuyler C. Cunningham, LICSW, LCSW-C, LCSW, Board Certified Diplomate, is an expert ADHD practitioner and award winning Social Worker. In addition to founding the Center for Neurocognitive Excellence, he was a Principal Investigator at the National Institutes of Health Clinical Center, chaired the Research Committee for the Social Work Department, and trained graduate Social Work students. Mr. Cunningham is a sought after speaker and has published on various aspects of mental health.

Tim Norman, LCSW, M.Ed is the founder and owner of the ATTN Center. As an expert in the field of ADHD, Tim gives lectures across the country, teaches workshops, develops ADHD programs and seminars, and co-wrote the book Helping Your Husband With ADHD. Before becoming a therapist, Tim was a public school teacher in the South Bronx, where he developed adolescent curriculum and standardized testing procedures that have been implemented state-wide.

NEXT STEPS

Now that you're done with this book, here are some next steps for you.

1. JOIN THE COMMUNITY

Join our community ADHD and Neurofeedback community and get helpful information about ADHD and Neurofeedback, including early access to upcoming books. Visit our website and follow us on social media for additional resources–articles, videos and events to support your journey.

www.adhdandneurofeedback.com

2. SHARE THIS BOOK

Please write a review on Amazon and tell others who you think will enjoy this book. Spreading the word helps to reach new readers, grow this movement and the continued production of similar content.
Write a review here:

Also, if you haven't yet read our other titles, now would be a good time to take a look.

www.adhdandneurofeedback.com

3. CONTACT US!

We are happy to consult with you about your case or your child's case. Visit our website for information on how to contact us directly.

THANK YOU

www.ingramcontent.com/pod-product-compliance
Ingram Content Group UK Ltd.
Pitfield, Milton Keynes, MK11 3LW, UK
UKHW021655190726
13853UKWH00001B/270